The
Two Monopole
Particle Universe

A RADICAL LOGICAL ALTERNATIVE
to the
STANDARD MODEL OF THE ATOM
the
BIG BANG
and
COSMIC INFLATION MODELS

TONY NORMAN MARSH

First published in paperback by
Michael Terence Publishing in 2024
www.mtp.agency

ISBN 9781800947399

Michael Terence
Publishing

Dedication

*I dedicate this book to my beautiful Wife,
my wonderful loving family
and the most wonderful Mum in the world.*

Preface

This book was inspired after watching a 'Royal Institution' lecture on quantum mechanics. Following this, I was shocked by how many problems remain outstanding in both quantum physics and cosmology today.

I am a 74-year-old independent theoretical particle / astrophysicist and successful inventor with patents granted. I have a background in both mechanical and electrical engineering, and worked for 11 years in the technical side of 'British Telecom', as a technical testing engineer trouble-shooting. I am also a 'Seer', often being able to see past problems that others can't, and have spent my life solving problems.

I decided to look closely into why these problems have persisted for around 100 years and to try to solve these problems from the perspective of a successful inventor, using logic not mathematics, which I blame for the crisis in physics and cosmology at the present time.

When you can get supposedly very well educated professors defending the ridiculous model of the 'BIG Bang' and 'Cosmic Inflation', where it is claimed that all the billions of billions of tons of matter that fills the whole of the universe, emerged from a tiny point smaller than an atom, and then inflated to fill the whole of the universe in less than a billionth of a billionth of a second, you have to ask, WHAT ARE THEY ON?

For the past two years, I have been trying to get a scientific paper accepted, but because I am not part of any academic institution, I have hit a brick wall. So, I have decided to turn the complete hypothesis into this book, for the benefit of anyone interested in the subject no matter what their level of academic knowledge.

I would hope that this book will inspire the next generation of physicists and cosmologists to look at things logically first, before being brainwashed by complicated mathematical equations and just accepting what they are told without questioning it. Yes, mathematics has a very important part to play, but not at the expense of logic. No matter how complicated or beautiful mathematical equations are, trying to prove the opposite, if someone drops a concrete block on you, you know it is going to hurt a bit!!! Like 'Bob Dylan' said in one of his songs, "You don't need a weatherman to tell you where the wind blows".

I have composed this hypothesis so the basic concept of what I am proposing can offer a logical and feasible alternative to the present accepted standard model. I very much hope that even the academic community will look objectively at what I propose alongside the mathematically dominated standard model, as I honestly believe that this Hypothesis has more credibility than the status quo. This book is aimed at everyone who has an interest in these subjects no matter how old they are, or how well educated they may be. I hope you enjoy this book, and if it makes you question what you are told, I will feel I have succeeded. Thank you and enjoy.

Tony Marsh

Contents

The Two Monopole Particle Universe

Sole Author: Tony Norman Marsh

In this hypothesis, I set out to present a radical alternative to the standard model of quantum mechanics and cosmology, based on logic as opposed to mathematics, which could explain the following:

Dark Matter: Dark Energy: Antimatter: cosmic nucleosynthesis: and two forces of Gravity.

A more rational model of the atom, where the relationship between the nucleus and the electron is fully explained:

The speed and ultimate distance that light and all electromagnetic radiation can travel:

An alternative explanation for the 'Cosmic Microwave Background radiation' and 'Redshift':

An explanation eliminating the 'Horizon Problem' without cosmic inflation:

An explanation for an infinite, nonexpanding universe that has always existed, without a 'Big Bang':

The cornerstone of my hypothesis is that everything that exists in the universe is composed of just two incredibly small monopole particles, which I contend are the only two true fundamental and finitely small particles that exist in the universe.

One is a negatively charged monopole particle called a 'Harveytron'.

And the other is a positively charged monopole particle called a 'Dannytron'. All other particles are composites of these two particles.

Background

Since there are so many inconsistencies and unanswered questions in the standard model of quantum physics and cosmology many physicists are sure that the standard model needs replacing, particularly in light of the data coming back from the 'James Webb Space Telescope' (JWST).

I have compiled this hypothesis, which I believe can answer many of the outstanding problems that exist in physics and cosmology at the present time. It is expected that it will provoke a great deal of debate, which is not a bad thing. On the plus side, I believe it will resolve many outstanding problems, such as the missing mass of the universe, the conflict between waves and particles, dark matter, dark energy, antimatter, ultimate velocity within the universe, energy transfer, the strong and weak nuclear forces and Gravity.

The end section is really a separate hypothesis, which proposes a logical alternative to the 'Big Bang Theory', and cosmic inflation, with a logical alternative explanation for what we see as the 'Cosmic microwave Background Radiation' and 'Redshift'. However, since it follows on from the 'Quantum Mechanical aspect', it does complete and complement the main body of this hypothesis.

It is accepted that the normal way of presenting challenges to the accepted standard model of QM and Cosmology, would be by mathematical equations. However, since I believe that many of the accepted values and parameters used to construct these equations are wrong, I have presented this hypothesis in a way that can be explained logically, will fit with the accepted proven laws of physics, right through from the very smallest sub-atomic particles to cosmology, and be understood by people with varying degrees of technical intellect.

Another reason why I have not included mathematical equations in this hypothesis is because what I propose is purely an hypothesis, and not a theory that has been backed up with empirical experiment.

Also, because at the present time the particles that I hypothesise have no quantitative dimensions, weights, or electromagnetic values associated with them. Therefore, mathematics would be a distraction from the basic ideas I am trying to put forward to be considered alongside the standard model.

Citations

I do not cite any scientific papers by various authors going back into history for the following reason.

As this hypothesis covers so many different areas of quantum mechanics and cosmology, and could be looked on as a 'Theory Of Everything' (TOE), I would have to cite hundreds of previous submissions that have been made on every element covered by this hypothesis.

Various elements of this hypothesis have been mentioned by various people in the past, but since this hypothesis is a radical alternative to the accepted model, it forms a unique model in its own right in how it all fits together. Since this hypothesis is all my own work, straight out of my head, I do not see the need for quoting the historical past, as I am not using other peoples' material, or adaptions thereof to present this hypothesis. Any previously submitted material is freely available to anyone interested in the subject offering a far greater level of accuracy than I would be able to provide.

The main consideration has to be the concept of what I am proposing, since this is a unique hypothesis, and not a variation on previously submitted work.

I am not aware of any hypothesis that has been proffered in the past that this hypothesis could be inferred or interpreted, as a copy of.

The problem with the subatomic world is that with our present measuring technology, we can't see or measure the makeup of the atom, so any model Is really only an educated guess, based on the information we have, plus our own intuition based on the physics we can measure, and logic. The one thing that can be accepted for classical physics and subatomic physics is the speed of light, with a caveat, that this can only be confirmed over relatively short time and distance scales.

My hypothesis is just the same, with the inability to see or measure the sub-atomic world. However, my hypothesis is able to unite quantum physics / mechanics with the classical physical world, explain dark energy, dark matter, antimatter, why the speed of light is what it is, and why this is the ultimate speed that anything can travel in the universe, how electromagnetic radiation and gravitational waves are able to travel through space, and gravity, logically. It is a model that could actually work. Any model, in the first instant, must be able to be explained logically without mathematics. It may also accord with the results coming out of the work at the Large Hadron Collider, (L.H.C). albeit for different reasons.

Classical quantum physics/ mechanics, which the standard model of physics is based on, describes the atom as consisting of a central nucleus, made up of positively charged 'Protons' and neutral 'Neutrons', these themselves made up of many different exotic particles, surrounded by orbiting negatively charged electrons some distance from the nucleus. The problem I have is this, the amount of empty space between the nucleus and the outer boundary of the atom is massive in quantum terms. Since the volume of the nucleus is around 100,000 times smaller than the volume of the whole atom, and the volume of the total number of electrons is 2,000 times smaller than the number of protons contained in the nucleus, the empty space within the atom does not make sense.

The latest widely accepted theory of the atom describes the whole of this empty space being occupied by an 'Electron Cloud', where the electrons or electron as in the case of the hydrogen atom, fly around at speed, being everywhere at the same time. This, if looked at logically, is an absurdity for so many reasons, not least the following.

If the electron was a solid particle, that was flying around the whole of the void of the atom, forming a cloud, which is not possible with a single particle, it would have to move at an incredibly fast speed and would have to change trajectory thousands of times a second. As it is a particle with mass, it would need energy to initiate and maintain its momentum. This would release heat and therefore all matter would be generating heat. This energy would have to come from, somewhere, where? Since all matter is plainly not emitting heat, the electron as described in the standard model cannot be moving, but must be at rest unless excited by an external force. If the electron is at rest, it would not have any angular momentum to stop it combining with the nucleus, as it is of opposite charge.

This problem is excused away by saying that the electron is both a particle and a wave, according to 'Schrodinger's' wave function, with no explanation of what the wave consists of or how the charge is distributed.

The standard model has the electron orbiting the nucleus with a charge equal but opposite that of the proton, so the proton is 1, and the electron is 1. This is derived by the combined charges of the components that the nucleus is made up of. Since according to the standard model, the proton and the neutron are each made up of three quarks, either two up quarks and one down quark, or two down quarks and one up quark, the sum of their fractional charges either add up to 1. in the case of the proton, or 0 in the case of the neutron.

Under beta decay, each of these quarks are able to transmute into their opposite quark, in the process emitting a neutrino, or anti-neutrino and an electron. This would imply that the nucleus contains 6 electrons in the case of a simple helium atom, and therefore 6 times the charge of the electron orbiting the nucleus, which would make the proton negative. This again proves the standard model wrong, and the mathematics flawed. Further to this, the mathematics for these charges, conveniently leave the nucleus as positive 1 and the electron at negative 1. There is no explanation for all of the other particles that make up the particle table, as to how these particles affect the charge of the nucleus.

The standard model's description of the atom has the nucleus made up of six quarks, which are said to be fundamental, and cannot be subdivided, but then it says that in beta decay, the quarks can eject an electron and a neutrino. If they can eject these particles, then they must contain these particles, therefore they are not fundamental. Like I mentioned earlier, where do all the other so-called particles that have supposedly been discovered, fit into the nucleus, which is only supposed to consist of six quarks? It is a complete mess, yet no one seems to question it.

Electrical charge

The second thing that doesn't make sense is the charge allocated to the electron. Since the negative charge of the electrons, is said to be equal to the positive charge of the protons contained within the nucleus, the charge carrying capacity per unit of mass of the electron, would have to be unbelievably large, considering how small the electrons are compared to the Protons. It would be like having a tiny watch battery, with the charge holding capacity of a large tractor battery or larger. Electrical transfer would not seem to be feasible, with so few electrons containing the whole of the allocated electrical charge, orbiting around the void of the atom.

It will be argued that although the proton is much larger than the electron, there is no reason why its charge density should not be much lower per unit of mass than the electron. This is a legitimate point, but in reality, one would expect a much closer charge density per unit mass to exist between the proton and electron.

Points like these highlight where the standard model fundamentally breaks down and the whole of quantum mechanics becomes flawed. It is also the reason why it is incompatible with standard physics.

To base a whole theory on an assumption that a point like solid particle can form a cloud and be in hundreds of thousands of different places at the same time, and have the same charge holding capacity, as something 2,000 times its mass, is frankly absurd and defies logic.

Since the electron can't be observed, and in the standard model the electron is given an estimated value for its weight and charge, the empty space within the atom with no weight allocated to it, could explain some of the inconsistency between the expected weight of the universe and the actual weight derived from mathematical calculations, albeit the parameters that these calculations are based on, may themselves be flawed.

The whole of quantum mechanics is only an assumption, which does not work as soon as the boundary of the atom is crossed into the classic physical world. The only way the mathematical equations can be solved is by introducing an 'uncertainty principle', 'probability theory', Superposition, and 'imaginary numbers'. In fact, none of the particles of the atom can be seen, therefore, it is only conjecture that the standard model is based on. The ever-increasing number of so-called new particles proposedly being discovered must lead one to question the logic behind them. Quite honestly, it is too complicated to be a feasible working model.

There are so many inconsistencies and anomalies in the standard model, some of which are as follows.

a) From standard physics, we know that for electricity to flow there needs to be a conducting medium for electrical transfer, such as a physical conductor in the case of conventional electricity flow, or an ionised gas in the case of static electricity. This must apply to the subatomic as well as the atomic world. With the empty space within the atom there is not enough conducting material for electrical transfer to pass from one atom to the next. This becomes obvious when we consider the transmission of electromagnetic radiation such as light, and also the transmission of gravitational waves, or should I say 'Shock Waves', through space.

In the atmosphere of the Earth, transfer of light could be proposed as being facilitated by the gases of the atmosphere conducting the electromagnetism, although I do not believe that this is how it is transmitted. However, beyond the atmosphere of earth, space is said to be a vacuum, which would contain no medium for conducting light, (Electromagnetic Radiation) or shock waves.

It will be argued that space is full of fields, but this is another cover all way out. What do these fields consist of? They would have to consist of something, otherwise space would be empty.

This adds weight to my hypothesis that there needs to be another particle that forms a cloud throughout the universe, to act as a conducting medium for all electromagnetic radiation and shock waves to travel through space. Without such a conducting medium, it would be impossible to see back in time or be able to view the 'Cosmic Microwave background radiation'.

I must clarify my perspective on so-called 'Gravitational waves'. I say "so-called", because in fact they have nothing to do with gravity, as they are in reality shock waves, they just move through a medium which I contend is what constitutes one of the two forces of gravity.

I find the notion of many different fields or strings highly questionable when looked at logically. It is far more likely that there exists just one medium that fits the requirements that I have mentioned. My 'Harveytron Cloud', fits these requirements exactly, and works logically.

In my hypothesis, there is no such thing as empty space, any space that is not a nucleus of an atom, or antimatter, is filled with the proposed new negatively charged monopole particles. Because of the way that I believe atoms are constructed, it is not possible to have or create a true vacuum. This is because, the particles I propose are so small, and there are no materials that would block their passage. Any attempt to remove these particles would fail, because, since every space is filled with them, the ones that were removed from inside any kind of vessel would immediately be replaced by the ones contained in the atoms making up the vessel. The ones depleted from the vessel, would be replaced from the 'Harveytron cloud'. Obviously, molecular matter, such as the gases are capable of being removed from a vessel, but not the negatively charged particles I propose.

b) Another indicator that points directly to the need for the cloud that I propose, and is missing from the standard model, is a medium to convey the long proposed gravitational waves (Shock waves). For a wave to propagate, it needs to have a medium in which to propagate, like waves through water; or soundwaves though air.

If space were to be a vacuum, then there is no way that light or shock waves can travel through it, therefore, we would not be able to see the sun or anything else in space and everything would be black.

c) The standard model of the atom has the nucleus being made up of many different types of particle that have proposedly been discovered in particle colliders, the number of which are growing on a regular basis. Since these particles can't be observed, they are only items of conjecture. The fact that any of the particles proposedly making up the nucleus decay within a fraction of a second, following a particle collision, must beg the question, are these real particles?

It seems quite a bizarre way to find out what something is made of, to smash nuclear material together at immense force and call the resulting detritus a new particle when it conforms to the dimensions and electrical characteristics that theoretical mathematics predicts.

If one were to smash thousands of cups together, it would not be surprising if every so often, a handle would turn up complete. This in the eyes of the particle physicists would be referred to as a new particle.

In classical physics, if we want to know how something works, or what it is made of, we can disassemble it and examine its component parts, which remain in their form and do not disappear. Why should it be any different for subatomic particles, and why should the laws of classical physics change just because the boundary of the atom is crossed?

It is my belief that all exotic particles, are in fact man-made particles, or should I say, pieces broken off of the atom, making the resulting mass unstable and probably radioactive for the period they exist prior to decaying and returning back to reform in the atom.

I believe that unless a new model is adopted, the solution to quantum physics / mechanics will never be resolved.

Another anomaly that the standard model does not explain, are 'glueons'. These are supposedly what glue the nucleus together, how? It is said to be the strong nuclear force, but there is no explanation as to what this force consists of, or how it works.

My hypothesis, explains exactly how the nucleus is held together tightly in a logical way that can be understood, and conforms to the known laws of physics. It also explains logically the strong and week nuclear forces, also conforming to classical physics.

Alternative Hypothesis

The Two Monopole Particle Universe

The cornerstone of this hypothesis is that everything that exists in the universe is composed of just two incredibly small monopole fundamental particles.

One being a negatively charged monopole particle called a 'Harveytron' which fills every available empty space throughout the universe in a cloud, and is what constitutes 'Dark Matter'. The negative force of repulsion that it produces is the 'Dark Energy' and is one of the two forces of gravity.

The second is a corresponding positively charged monopole particle called a 'Dannytron which is only found in nuclei of baryonic and 'Antimatter', and is the other force of gravity.

The Electron

It is proposed that the accepted view, that an electron is an elementary solid particle, is incorrect, and that it is in fact a collection of the much smaller negatively charged monopole particles that I have just described as 'Harveytrons'. I believe that these particles would be the finitely smallest particles in the universe, and would constitute everything that exists in the universe, with the exception of the positively charge component contained in every nucleus that exists, including the embryonic nuclei of antimatter. Also I propose that these particles would form a cloud, the 'Harveytron' cloud, that fills every available empty space within the atom and the universe.

These negatively charged monopole 'Harveytron' particles are 'Dark Matter Particles', and permeate the whole of the universe.

Just to emphasise the point, I believe that the electron, as a solid particle does not exist, and my model gives a very good logical explanation for the latest accepted view of quantum physicists that there exists an 'Electron Cloud', although there is no explanation in the standard model of how the limited number of so-called solid electrons can form this cloud.

It is proposed that the electron as it is described in the standard model, is in fact a parcel of the much smaller negatively charged monopole 'Harveytron' particles described in my hypothesis, and is the quantity of negatively charged particles to be contained together, before the charge they contain is released as a Photon, or as a pulse of energy outside the spectrum of visible light. This quantity of charge is a constant, throughout the whole of the universe, and is the quantity of charge contained in a 'Quanta' of light or electromagnetic radiation, when the threshold quantity of charge able to be held within the boundary of the atom is reached.

This is analogous to an electrical circuit, containing a power source, a capacitor, and an SCR, whereby, when power is applied to the circuit, the capacitor begins to charge, until the trigger voltage of the SCR is reached and It fires, at which point the capacitor discharges in a pulse.

The way that these clusters form, is due to the interaction between the opposing positive and negative forces in the nucleus and any external force applied to the atom pushing the negative particles up to the point where the pressure cannot be contained by the atom, and the energy contained in these bunched particles is released to the adjoining atom. The quantity of negative charge surrounding each nucleus, is strongest nearest the nucleus, and gets weaker the greater the distance from the nucleus. At the point where the force of attraction to this charge from the nucleus reaches equilibrium with the force of repulsion between the negatively charged monopoles is reached, this is the boundary of the atom. This is also the boundary that the positive charge forming the strong nuclear force extends to. From this point out, there only exists the negative force of repulsion in every direction in the 'Harveytron Cloud'.

I contend that the whole of the available empty space within the boundary of the atom, as described in the standard model, is filled with these particles, and the amount of negative electrical charge equal and opposite to the positive charge carried by the nucleus, attributed to the electrons, as described in the standard model, is distributed throughout the mass of these particles.

The negative electrical charge that would be carried by the electron in the standard model is in fact, just a small proportion of these particles, and of the negative charge within the area surrounding the nucleus and the outer boundary of the atom. I also hypothesise that these particles do not fly around at speed, but are at rest, unless exited by an external force, other than trying to get closer to the positively charged particles contained in the nucleus, and contributing to the strong nuclear force.

Beyond the last layer of the nucleus containing the positively charged monopole 'Dannytron' particles forming the nucleus, as I will describe later, there are only the negatively charged monopole 'Harveytron' particles in a cloud that encompasses every space within the universe. This cloud is negatively charged, and forms a negative force of repulsion exerting a repulsive force in every direction.

As these negatively charged particles cluster around the nucleus, they are held very strongly by the positively charged particles making up the nucleus. As the positive and negative particles are monopoles that do not give up their force of attraction and repulsion, there exists a complex interaction between these particles in the nucleus. The attractive and repulsive force of each particle extends through adjoining particles. Although the negatively charged particles are trying to repel each other, this repulsion is a standard pressure in the 'Harvetron Cloud'. However, because the positively charged particles are trying to attract the negative particles from every direction, and their force of attraction extends beyond the boundary of their neighbouring oppositely charged particles, there is in effect, a double force pulling and pushing the nucleus extremely tightly together. This is the 'Strong Nuclear Force'.

As more negatively charged particles try to get to the nucleus, a dense shell of the negatively charged particles forms due to the very strong attraction of the positive particles in the nucleus. As these negatively charged monopole particles can't get any closer to the nucleus, this is the boundary of the atom. From this point out, there exists only the negative force of repulsion, and equilibrium is reached in the 'Harveytron Cloud'.

The negative force of attraction to the nucleus gets weaker the greater the distance from the nucleus, until the force reaches equilibrium within the cloud, to become the weak force, and a constituent of gravity. It is probably this force that is referred to as the 'cosmological constant'.

At the point where the boundary of the atom would be defined, it is possible that the negatively charged monopole particles, might bunch up as they are being pulled by the positive particles in the nucleus, but also being repelled by the negative particles already in this region. This could be what is visualised as an electron. Any applied force would then cause the energy contained in these bunches, to be released as a photon.

The Photon

In my model, I see there could be two possibilities of what the photon is composed of. One is the cluster of the negatively charged 'Harveytron' particles I describe, or a single positively charged monopole 'Dannytron' particle encompassed by the 'Harveytron' particles. This second option would counter the argument as to "why doesn't the cluster of the negatively charged 'Harvetron' particles merge into the 'Harveytron Cloud'," which would be a valid point.

From what has been described, it will be seen that there is a logical explanation as to the relationship between the positive charge of the nucleus and the negative charge surrounding it. This completely eliminates the concept that the electron is not allowed to fall to the nucleus, as stated in the accepted model of quantum mechanics, even though it is supposed to have an equal but opposite charge to the nucleus.

This would accord with the idea of an electron cloud, and with the pictures taken of atoms, where there appears to be a fuzzy haze surrounding the nucleus. I contend that instead of electrons whizzing around the nucleus, the phenomenon of an electron is created from this cloud, in discrete quantities, (Quanta).

Atomic Reactions

Very briefly, I will just touch on this subject. Looked at logically, the emission of a photon, or co-valent bonding, could be looked on as the lowest form of atomic reactions. Anything being emitted from an atom is 'Fission' and anything being received by an atom is 'Fusion'. This is a subject that is very much open to research.

The Nucleus

I would also hypothesise that the nucleus of the atom is radically different from the standard model. I believe that the nucleus is made up of just two types of particle, the small negatively charged monopole particles that I have described as 'Harveytrons', and The positively charged monopole particles, that I described as 'Dannytrons'. These particles may well be dimensionally the same, but of opposite charge. They are also monopole magnets, which retain their forces of attraction and repulsion even when surrounded by their oppositely charged particles. It is this fact that binds the nucleus so tightly together.

I believe the makeup of the nucleus would be either like a 'Gobstopper' sweet, starting with a positively charged particle, which is encompassed with negatively charged particles, then successive layers of alternately charged particles in shells, or as just a mixture of these two particles. The different elements being defined by the amount of positively charged particles contained in the nucleus of each atom.

From this description, it will be realised that the positively charged 'Dannytrons' will never be found in isolation, but always surrounded by the negatively charged 'Harveytrons'. However, the 'Harveytrons' do exist alone in the 'Harveytron cloud'.

Whichever way the Nucleus may be composed, the result would still be the same. All of the particles making up the standard model's table of particles are composites of these two monopole particles.

As described above, there is presented a logical model of the atom, where there is proffered a logical explanation of how the positive and negative forces pertaining to the atom act upon each other. This accords with actual observational evidence, without the need for any ambiguity, such as 'Uncertainty Principle', 'Superposition', 'The Wave Function', 'Probability Theory', 'Exclusion Principle', and the like.

Elements, Isotopes, chemistry

My model of the atom would still respond the same as in the standard model.

The different element are determined by the quantity of the positively charged 'Dannytron Particles' contained in the nucleus. The larger the number, the higher the atomic number element it is.

As the quantity of these particles contained in the nucleus are not constrained by any given quantity, such as a whole proton or neutron, as in the standard model. The variation is infinite, as the particles are so small. There may be thousands of these particles that would make up the equivalent quantity of positive charge as one proton. This explains the capacity for Isotopes to exist.

Chemical bonding

This works the same as in the standard model of the atom, the only difference, is that instead of a single solid electron forming a co-valent bond, a cluster of my 'Harveytron' particles, or the energy they contain is what moves.

The fact that there exists a threshold level in the outer boundary of the atom that has to be reached before a quanta of energy is released, means that there is a potential deficit or excess about a mean value of charge contained in the outer boundary of the atom, allowing it to receive or eject energy from or to other atoms, in discrete packages, (Quanta of Energy).

Beyond the nucleus, there are only negatively charged particles in a cloud around the nucleus to the boundary of the atom and then on throughout the whole universe. When electrical energy and light moves from one atom to another, as depicted in the standard model, only part of this charge leaves the atom as a package, as would be attributed to an electron movement described in the standard model, producing a 'Photon'. The deficiency of negative charge being replaced by the negatively charged energy exerted on the atom to dislodge the package of charged particles from the atom. The existence of these particles forming the cloud, could explain the inconsistency between the expected estimated mass and the actual missing mass from the universe.

I do not believe that there are any electron movements as described in the standard model, where electrons jump between higher and lower orbital rings of the atom. I believe that it is only the 'Harveytron' particles that I propose occupying the outer boundary of the atom that are responsible for the transmission of energy from one atom to the other and where molecular bonding takes place.

As far as I am aware, my hypothesis would not conflict with the known and proven accepted laws of physics, but may aid in the resolution of problems as yet unresolved. However, I would expect it to be in conflict with some assumed accepted facts derived purely from manipulated mathematical formulae. It is also very possible that my model could accord with the work that particle physicists have been doing for a number of years.

By this I mean, even though my model of the nucleus is different from that of the standard model, from my description, it can be seen that if nuclei of my model were collided or parts of it, the resulting detritus, would consist of many fragments of atomic matter of varying dimensions, and electrical characteristics. This may accord with the particles that make up the standard model.

Depending on the proportion of positive and negative particles contained in these pieces, may account for the phenomenon referred to as spin in the standard model. Also it could explain why in violent reactions, Hydrogen is formed, when the right quantity and mixture of these two particles are broken off from atoms. This description strengthens my contention that the so-called particles making up the table of particles in the standard model are not actual particles but detritus.

However, I stand by my hypothesis, and feel it has more merit to be a more workable and logical model that fits well with our known and proven classical laws of physics. It also provides a workable and logical explanation for the unanswered questions of quantum mechanics and cosmology today.

Light or energy

I propose that the speed of light is actually determined by the speed at which energy can move from one of these particles to the next, whether this be by individual particles or- quanta. This obviously doesn't alter the speed at which light travels, but gives a reason why the speed of light is what it is, and why this is the ultimate speed that anything can move throughout the universe.

Light and all electromagnetic radiation, are quanta of energy contained in sub-atomic particles, and most likely can travel without being subject to friction, due to the negative force of repulsion between the particles of the 'Harveytron cloud'. Anything else, is composed of whole atoms, which is subject to gravity, restricting its freedom to travel as fast. Also, any nuclear matter will contain the positively charged 'Dannytron' particles, so there will be an element of drag due to the attractive nature of these particles to the 'Harveytron' cloud.

The reason why light can travel so fast is because it is either these 'Harvetron' particles, or energy, being transferred from one particle to the next in a bunch (Quanta). Because there exists a force of repulsion between these particles, it may be that the energy that is released is not subject to friction. Also, I hypothesise that the pulse of the particles or energy does not actually travel over any distance, but is transferred from one particle to the next. Energy transfer through the 'Harveytron cloud', would be by a cluster or quanta of these negatively charged particles, releasing their energy by an applied force, strong enough to overcome the threshold level, and eject the package of energy it contains, the same way as a photon, as depicted in the standard model. I believe that the amount of these particles within the package would always be the same size, and may turn out to be a new constant.

It may be that it is the cluster of these particles together with the charge they contain that passes from atom to atom, and because they are all negatively charged and repelling each other, they are not touching, so there is no friction encountered.

Because the distance they travel is so small, the movement across the distance of one particle instantaneously moves many more particles. If you arrange a line of ball bearings over a given distance, all touching each other, then apply a force enough to move the first ball bearing the distance of one ball, the last ball moves instantaneously. If there were one hundred balls in the line, then the time taken for the effect of the applied force to affect the last ball, would be one hundred times quicker than if the first ball had been required to travel the combined distance of the one hundred balls.

Energy

I hypothesise that the two monopole particles I propose, are the only fundamental elementary particles that exist, and are what we refer to as pure energy.

These particles cannot be destroyed, they can only be changed into matter of some sort, whether it be 'Baryonic matter', 'Antimatter', or 'Dark Matter'. I would contend that above a certain temperature, baryonic matter would disassociate into free positive and negative particles as their magnetism reduces. In their disassociated state, they constitute 'Plasma'.

Electromagnetic radiation

I believe that there exists a misconception with regard to electromagnetic radiation, which may have resulted from reliance on equations using the wave function.

The general conception seems to be that electromagnetic radiation travels as waves. It is my belief that radiation is emitted from atoms as quanta of energy, as a ray in a straight line, not as a wave. The frequency, is the number of these quanta being emitted in a measurement of time, in our case the second. What is referred to as the 'Wave Length', is the distance between these quant, the frequency, is the number of these quanta passing a given point in a given time, there is no amplitude, and therefore no wave.

Types of matter

In my hypothesis, there are only two types of matter, positively charged monopole 'Dannytron' particle matter, as contained in all of the nuclei in the universe and the negatively charged monopole 'Harveytron' particle matter, making up around half of each nucleus, a percentage of antimatter embryonic nuclei, and the 'Dark Matter' 'Harveytron cloud'.

Gravity

As we know, like charge repels. From what I am proposing, it will be realised that there will exist a massive excess of negative charge throughout the universe. This would result in a force of repulsion between all of these negatively charged particles trying to force each other apart in every direction, creating a negative pressure throughout the universe. Once the potential difference between the negative charge of the cloud and the positive charge of the nuclei have been equalised, there will still remain a huge excess of negative charge. The force of this charge, is the 'Dark Energy', and this 'Dark Energy' force of repulsion, is one constituent of gravity, which I believe is a force of both attraction and repulsion in an interaction similar to that of the atom that I have described.

The force of attraction is for the positively charged particles contained within the nuclei of all of the atoms in the universe, such as the planets and any physical observable nuclear matter in the universe. Once all of this matter is surrounded by these negatively charged particles, there would exist just a force of repulsion trying to force each piece of physical matter apart, this force is what keeps all physical matter suspended in space, like a ball bearing can be held in suspension in a magnetic field, and is one of the forces of gravity.

Since we are made up of atoms, this force acts on us the same as all matter, and is what keeps us anchored to the earth.

The other component of gravity, is the force of attraction between the 'Harveytron Cloud', and the positive charge of the 'Dannytron' particles contained in the nuclei of atomic matter.

The reason why quantum mechanics and gravity do not accord with classical physics is because half of the data is missing from calculations being done. This is why the force of gravity is thought to be so weak, as it is only the force of attraction that is being considered, which is only half of the picture.

Looked at logically, to think of gravity as just a force of attraction doesn't make sense with regard to observed phenomena, such as what stops planets and galaxies from merging together in space, or why the gasses in our atmosphere do not fall to earth, and the like.

Possible proof of repulsive force

If gravity were just a force of attraction, when a drop of liquid is dropped into a volume of liquid, it would be held tightly and just spread out and merge. But it does not do this, it rebounds back off of the surface, and as it rises up, there is a column of liquid tapering at the top but the point is pinched off forming a perfect sphere. This does not accord with just a force of attraction, but would accord with the type of gravity that I describe, with the 'Harveytron' cloud pushing in every direction, pushing the repulsed drop upwards forming the drop. As the force of attraction is slightly stronger than the force of repulsion, the drop and the column of liquid eventually combines with the bulk of the liquid.

In cosmology, if gravity were just a force of attraction, it would be hard to explain what stops planets and galaxies from merging into one solid mass. In rotating galaxies this can be explained by the interacting forces of centrifugal force and the attractive force of gravity.

However, this would not explain why galaxies do not combine, particularly those that are parallel with each other. The only logical explanation is to have a repulsive force in the universe as I am proposing.

A very simple experiment to show a repulsive force is as follows. Take two pieces of A4 paper that are completely flat and not under any stress from bending. Cut a strip lengthways from the pair about 1 ¼ inches wide. Cut 3 or 4 smaller pieces from a similar strip to form squares, and place these between the two long strips at one end. Make a hole through the middle of the square of this top section, through which you can pass a suitable thin rod that allows the strips to swing freely. Make sure that the top pieces are pushed tight together, so that the strips hang freely down parallel to each other. Place these in a long glass vessel, such as a spaghetti jar, so that they are not influenced by air currents, cover if you think appropriate. You will notice that the gap between the two strips is slightly larger at the bottom than at the top.

If the spacer between the two strips is too small, the two strips will be pulled and pushed together.

This I believe shows that there does exist the negative force of repulsion acting in every direction due to the existence of proposed 'Dark Matter' 'Harveytron' cloud producing the 'Dark Energy' repulsive force. I feel it also proves the existence of the two forces of gravity that I propose, where atomic matter containing the positively charged monopole 'Dannytron' particles are attracted to each other if the distance between two atomic masses is too small.

From this simple, logical experiment that shows an action, and an explanation as to why the action is happening, it will be seen that it will be possible to collate all of the data, with regard to surface area of the paper strips, their weight and distance apart etc, and devise the mathematical equations to complement the observations.

Particles and Waves

I can't see any conflict between particles and waves. All waves, are just particles being disturbed, so all waves are particles. I contend that for a wave to propagate and travel, it has to have a medium to propagate in. With the existence of the 'Harveytron cloud', there is provided just such a medium in which waves can propagate. I believe that there are just particles and energy comprising everything in the universe. All waves are just particles on mass oscillating to a given frequency. I further contend that there is no empty space in the universe, there are only physical objects such as planets and the like (Baryonic nuclear matter), Embryonic baryonic matter (Antimatter), and the rest of the space is taken up with the negatively charged monopole particles that I propose (Dark Matter). There are no weird phenomena in quantum physics, it's just that we have been trying to make sense of quantum physics without having all of the pieces of the puzzle to be able to complete the picture. I believe that the standard model is like a puzzle being put together, but the fundamental problem is that right at the beginning, one of the pieces although having the right shape to fit, and looked like the picture being constructed, it was the wrong piece in the wrong place, added to the fact there have been pieces missing.

No matter how many fancy equations are constructed to try to prove the picture right, without this piece being changed, the puzzle will never be solved. I believe my monopole particles are the pieces that are missing, because my model can unify quantum mechanics with classical physics, and can be explained logically.

The particles that I propose, explains the missing monopoles that are predicted should exist. The reason why they are thought to not exist as they have not been detected, is because for one thing they are so incredibly small, and the other is that everything that exists in the universe is composed of them.

I am firmly in the 'Einstein' camp when it comes to entanglement, I believe he is correct that a rational explanation should be possible to answer the entanglement question logically. I believe that in my model, this may be possible, because when a so-called particle is fired in the experiments, it is not a single particle that is being fired, but a cluster of particles that are already entangled. Also, any particle or cluster of particles being fired in the 'double slit experiment', are moving through the 'Harveytron' cloud, which would make any calculations invalid, as any measuring equipment or measuring material, is made of the same atomic material as that which is being measured. In other words, there would be a cluster of particles traveling through a cloud of the same particles, which would form a ripple of waves as they approach the double slit, this would obviously cause an interference pattern the other side of the slits.

Any measuring equipment would be emitting a stream of the same particles, which would have an effect on the original pulse, yet the original pulse of particles would in most cases reach the fluorescent screen, and depending on the direction of the observation beam, it will influence where on the screen this pulse will be detected.

It would be interesting, to put another fluorescent screen in line with the observation beam, to see what patterns form on that.

With what I have described, it will be seen that it would be like pushing a jet of water towards the double slit, completely submerged in water.

Because of the infinitesimally small size of the particles that I propose, the unit of charge that they hold is also incredibly small, but in bulk, it is of sufficient force to hold all matter in suspension. This is the week nuclear force.

Because there is such a massive in-balance between the positive and negatively charged particles, each negatively charged particle would be trying to get to the positively charged particles from all directions, the positive particles are forced together, forming a strong positive mass.

Since the positive mass is forced together into the solid matter in the planetary universe, it could be argued that there is a force of attraction of other matter towards a larger mass. This may be so if masses become too close to each other, but as there is still much more negative charge throughout the universe than positive, there would still be more of a force of repulsion than attraction in the equilibrium of the 'Harveytron' cloud. However, since the nuclear matter is being forced together so strongly, nuclear matter will have a larger charge per unit of area than that of the 'Harveytron' Cloud.

This situation results in the amalgamation of matter with matter dependant on the ratio between the mass of the matter, and the distance between each mass of matter. Since there is such an excess of negative charge, this could explain the slow expansion of the universe (if it is). I do not believe that the universe is expanding exponentially, in fact I do not believe it is expanding at all.

I hypothesise that quantum gravity and cosmological gravity, are one and the same interaction between the two positive and negatively charge monopole particles.

In the case of gravity as we normally understand it, this is how I believe it works. From what I have described, it will be seen that beyond the positive attraction of the nuclei to the

'Harveytron cloud', surrounding each atomic nucleus forming atomic matter, there will be a zone surrounding the physical matter such as the planets etc, where there will be a slightly stronger force of attraction to the 'Harveytron cloud' due to a slight excess of positively charged 'Dannytron' monopole particles contained in the atomic nuclei of physical matter (including all gasses).

Beyond this zone, the 'Harveytron cloud' reaches equilibrium, and a steady 'Dark Energy' force of repulsion is maintained, keeping cosmological matter suspended in the 'Dark Matter Harvetron cloud'.

Obviously, the repulsive force of gravity pushing matter apart, is also pushing matter close to other matter when the distance between matter is too close, as was demonstrated in the simple paper experiment I described earlier. This distance is proportional to the relative masses separated. Mass such as ourselves on earth, is held tightly to the earth by the positive force of attraction of all of the nuclei making up the mass, and the negative force of repulsion of the 'Harveytron' cloud. This is of a constant pressure, whereas the attractive force of the nuclei of atomic matter increases with mass. This is why, small mass is drawn to larger mass should it come too close for the repulsive force to keep the two masses apart, at which point the two masses merge. This was demonstrated by the small experiment I described, where, when the distance between the two strips of paper was enough to allow enough of the 'Harveytron' 'Dark Energy' cloud to act against the inner faces of the strips of paper, they were pushed apart further at the bottom. However, when they got too close, the two strips were forced together.

I hypothesise that all forces in the universe ultimately reduce to one electromagnetic force.

My hypothesis could have big implications for the LHC, because it may be possible to detect these particles with the existing equipment, as opposed to looking for the larger proposed particles that require a larger collider. I would not be surprised if they haven't already been seen in the so-called 'Crap' that is emitted following proton collisions. As I have said, I believe that all of the so-called exotic particles proposedly discovered forming the nucleus of the atom as proposed in the standard model, may not actually exist.

Antimatter

I hypothesise that the antimatter that has been detected, is predominately man-made, and constitutes matter that contains too few of the positive particles I describe to form a complete hydrogen atom. This is the type of matter that exists in cosmological plasma, before it forms into hydrogen on cooling following violent events such as supernova explosions.

The reason why generally we can't detect it and there is so little of it, is not because it has been annihilated, but that as the plasma cools and the two particles that I propose form, they combine until enough of the positively charged monopole 'Dannyton' particles have been encompassed by the negatively charge monopole 'Harveytrons' making up the 'Harveytron Cloud' to make the Hydrogen atom. At this point, the 'Antimatter' ceases to exist. This explains why the first atoms to exist in the universe are hydrogen atoms. I contend that all of the atoms in the universe did not appear at the so- called big bang or a period after, but they are being produced all of the time in the violent events I have just mentioned, and the so-called big bang, did not happen. This will be explained in detail later, along with nucleosynthesis. Recent evidence from the JWST, would tend to suggest that this idea is correct.

As nuclear matter is collided in colliders such as the LHC, it is very possible that some of the resulting pieces are of the correct ratio of positive and negative particles to be the antimatter that I have just described.

The fact that all of the so-called particles, that result from collider collisions decay in a fraction of a second, shows a propensity for matter to want to aggregate together back to their original states. In the case of the two particles I propose, it will be realised that it is not possible for the positively charged 'Dannytron' particles to exist on their own, as they will immediately be surrounded by the negatively charged 'Harveytron' particles.

In a collider collision, an artificial state of affairs arises, whereby fragments of nuclear matter are produced, that have an imbalance of positive and negatively charged particles, that renders them unstable. Fragments of varying sizes will have different electromagnetic characteristics, which would probably explain the ever increasing variety of so-called particles proposedly being discovered.

Unification

I think the main reason why the laws of standard physics differ from those of quantum physics, is that in quantum physics, only energy moves, or the particles that I propose surrounded in energy. Because these tiny particles, which are matter, have this energy field around them, they can move through the cloud without experiencing any friction.

Whereas, in standard physics, it is physical molecular matter that moves, which under the additional influence of positively charged attractive gravity, is subject to friction because of the attractive force of gravity pulling molecular matter towards earth.

Cosmology

My hypothesis is like all things in nature, it has a cycle, like subduction of the Earth's mantle returning to the centre of the earth, re-melted and expelled through volcanoes. The water cycle, the carbon cycle and so on.

I believe that the universe has this negative pressure and this pressure is maintained in equilibrium with a symbiotic relationship between violent reactions, such as supernova, producing new matter from that which has been converted into plasma and the consumption of matter by black holes (If indeed they do actually consume matter).

I also hypothesise, that everything in our known universe, did not emerge from nothing in a big bang, which defies logic and is quite honestly ridiculous. I believe that before the so-called big bang event, the universe existed and does exist, much the same as it is now, and extends to infinity as we understand it.

Summary

The above is just a brief description of my hypothesis, and it will be obvious that all of the points I raise can be extrapolated on a great deal. However, I feel it gives a good insight into a feasible and logical alternative to the standard model of quantum physics that could work better than the existing model.

Proffered by: Tony Norman Marsh

The Big Bang Misconception

Sole Author: Tony Norman Marsh

Abstract

An hypothesis to challenge the idea of the 'Big Bang', being how our universe began, and the idea that the universe is expanding. It is further proposed an alternative explanation for the 'Cosmic Microwave Background Radiation' (CMB), and 'Redshift'.

It is proposed that light and all electromagnetic radiation may have a limit to how far it can travel through space. It is also proposed that the universe did not emerge from nothing, but is a dynamic system, in which matter is constantly being created and consumed in a cycle over billions of years.

It is proposed that the universe as we observe it has always existed much as it is today, and extends to infinity.

It is further proposed that the so-called 'Big Bang' was not some kind of massive explosion, but is a misconception of what is being interpreted as the 'Cosmic Microwave Background Radiation'. The 'CMB' is the only proof offered for the 'Big Bang theory', and this can only be held up by an even more preposterous idea of 'Cosmic inflation', inflating the universe from a point of singularity to fill the whole of space in a minute fraction of a billionth of a second.

Background

Big Bang Theory

The current widely accepted view of how the universe came into being is the 'Big Bang theory', where from a sub-atomic singularity, an immensely hot dense plasma emerged. It is said that this tiny speck of plasma, then inflated to fill the whole of the universe in a fraction of a billionth of a second. From then on, the universe is said to have expanded at an exponential rate over 13.8 billion years to the present date.

It is estimated that the universe now extends to around 46 billion light years across, but the accepted view of the cosmological community is that the 'Big Bang' was at 13.8 billion years ago, which is as far back as we can see at the present time. The difference being attributed to cosmic expansion, this means that there are 9.2 billion light years of the universe beyond that which we can see.

Even though this is the accepted view, there are many problems with this model, in that observations do not match up with predictions, which are becoming increasingly evident by data emerging from the 'James Webb Space Telescope'.

It is a mystery why early galaxies are smoother than predicted.

It is a mystery why the universe is so homogenous.

It is a mystery why there is thermal equilibrium throughout the universe without causal contact, which is referred to as the 'Horizon problem'. The way that thermal equilibrium is explained, is by the theory of cosmic inflation, which, if looked at logically, is physically impossible, according to our known laws of physics.

The theory has no answer to Dark Matter, Dark Energy, antimatter or gravity.

Cosmic Microwave Background Radiation

According to the standard model of cosmology, the 'CMB' is the background radiation that was emitted at the 'Big Bang' and is the heat left over which was spread out throughout the universe.

This is a contradiction in terms, as inflation is said to have happened to dissipate the heat evenly throughout the universe. If the hypothetical phenomenal heat of this unbelievably small singularity was spread out throughout the universe in a fraction of a billionth of a second, there would be no heat left over, because it had all been dissipated to produce the thermal equilibrium that is observed today.

In the relativistic cosmological model, it is explained that the reason why the CMB can be viewed at the same distance in every direction, is because the 'Cosmic Microwave Background Radiation', is said to exist throughout the universe, due to the heat of the plasma that existed at the 'Big Bang', being distributed evenly during the fraction of a second inflationary period following the 'Big Bang'.

Looking at it logically, it is impossible for the 'Big Bang' theory to be correct, for the following reasons.

The theory is based on two very unlikely if not impossible assumptions. The first is that it is possible for all of the billions of tons of matter that makes up all of the matter in the universe, to be contained in a singularity smaller than an atom. The second is that it is possible that this tiny hot dense singularity could inflate to the size of the universe in a fraction of a billionth of a second.

This led me to look at the problem from another perspective logically, and to propose the following hypothesis, for an alternative to the 'Big Bang Theory', that simply and logically describes a model of how the universe works, without the need for 'Cosmic Inflation', and why the universe may be static and not expanding.

Alternative Hypothesis

The Big Bang

It is hypothesised that the 'Big Bang theory', is fundamentally wrong, and can be disproved conclusively by the following observations.

The cornerstone of the big bang theory is the phenomenon referred to as the 'Cosmic Microwave Background Radiation'. This is said to be the afterglow of the heat left over from the big bang, which is claimed to prove it happened.

However, the fact that the phenomenon called the 'CMB' can be viewed at the same distance 13.8 billion light years away, in every direction, leads to the following. The first conclusion that this would suggest is this, the only way that this could be true, is if we were at the centre of the universe, which is unbelievably unlikely.

For the theory to have any credibility, our galaxy should exist at a point anywhere other than the centre. As such, if our galaxy were to exist near the edge of the universe the distance to the CMB should be different in every direction, and not the same.

The Relativistic cosmological model

The counter explanation to this view is the relativistic cosmological model. In this model, it is stated that because the hot dense plasma that emerged from the singularity at the moment of the big bang inflated to fill the whole of the universe, we see back to it now at the same distance, no matter where we are in the universe.

This explanation when examined closely has major flaws. In the first place, this theory is totally dependent on cosmic inflation, which itself goes totally against all of the known laws of physics. In particular, all the matter that exists in the universe would have had to have travelled, not just faster than light, but many times faster than the speed of light. Also, if the radiation was spread throughout the universe by cosmic inflation, it then must be present throughout the universe, and not show up as a phenomenon at a distance of 13.8 billion light years away, it is a nonsense.

If inflation and further expansion has caused the big bang universe to now be at 46 billion light years across, this could mean that earth could be situated anywhere in this area.

As such, if our earth was situated right at the edge of this universe, there would have to be a different view towards the edge than towards the middle.

This leads me to postulate that what is being observed, must be due to some other phenomena related to electromagnetic radiation that produces the picture that is termed the CMB, at a distance 13-14 billion light years away.

I hypothesise, that wherever you are in the universe, even beyond our visible universe, the same view will show the phenomena we call the CMB in every direction, over the same distance, using the same equipment. This I would propose, is that the universe as far as we can observe it, has always existed much as it is now, and beyond what we can see is likely to be the same to infinity as far as we can comprehend it. As I will explain later, I also contend that the universe is static, and not expanding.

This presents us with a dilemma, in that we need to find an answer to what is causing what we have been observing and calling the CMB. From what I have described, the CMB can't be the afterglow of any form of big bang, so new research will be needed to find a solution. Two possibilities come to mind and I hypothesise the following.

It has always been assumed, that light and all electromagnetic radiation, can travel to infinity, and at a constant speed. This has always been my understanding, but now that may need to be reconsidered.

Distance and speed of light

One possibility is that light, and all electromagnetic radiation has a finite limit to the distance it can travel, and when it reaches that limit, it diminishes into the fabric of space. This I would propose is due to light and all electromagnetic radiation, losing speed and energy over time and distance travelled. This is very possible as we can only measure the speed of electromagnetic radiation over relatively short distance and time frames in cosmological terms. By this I mean, the establishment of the accepted speed of electromagnetic radiation, has been ascertained by light being reflected back from a distant object in our own galaxy, over a period of minutes.

However, when it comes to this radiation that has travelled billions of years, we can only calculate based on perceived parameters, that is to say, all calculations to do with distance, temperature, time and the like are based on the assumption that all electromagnetic radiation travels at a constant speed, which is accepted as being 186,000 miles per second, to infinity.

I hypothesise that in fact, electromagnetic radiation loses energy and speed proportional to distance and time, and as such, would impact on our perceptions of what we see and calculate.

It would only take a very small drop in energy and speed, to make the calculations we make wildly wrong, and our perception of what we see.

This would provide a very good argument to support the idea that the universe is not expanding, as it could explain the 'Hubble Redshift', which is the accepted explanation supporting the expansion of the universe.

Alternative explanation for Redshift

Redshift is said to be the stretching of the wavelength of electromagnetic radiation, due to the expansion of the universe.

I hypothesise that the universe is not expanding, and the explanation why the electromagnetic radiation moves to the red end of the spectrum over distance and time, is that the speed of electromagnetic radiation drops proportional to the distance it travels.

As electromagnetic radiation consists of pulses (quanta), the frequency is the number of these quanta that pass a given point in a given time, and the time we use is one second. This number is proportional to the distance travelled in one second of time, and it is universally accepted that this is around 186,000 miles per second.

Depending on how fast speed and energy are lost from electromagnetic radiation, over time and distance, relative to its original frequency and energy level, would determine what part of the spectrum we would see the radiation at any given time and distance it has travelled, without the need for cosmic expansion.

I propose that over billions of years, speed and energy are lost from this radiation, and the speed at which it travels, decreases proportionally in line with the degree of redshift that is attributed to the expansion of the universe.

This means that the number of quanta passing a given point in a given time reduces proportional to the distance travelled and the reduction in speed. Although the distance between each quanta remains the same as when the radiation was originally emitted, its position in the spectrum moves to the lower end because the number of quanta passing a given point in a given time, matches a lower frequency radiation emitted in a more recent period of time. This is due to the drop in speed, and not to the expansion of space.

If one considers that if EMR is initially travelling at 186,000 miles per second, it would only have to drop by 1 mile per second every 140,000 years, to account for the redshift we observe over approximately 14 billion years.

This miniscule drop in speed, over 140,000 years, and the phenomenal distance it travels, is very rational, and extremely logical. This would also accord with observations and overcome the problems associated with the 'Big Bang', and 'Cosmic inflation'.

This is a far more logical explanation for redshift, as it answers the question as to why the 'Cosmological constant' vacuum pressure Is maintained, where in an expanding universe, there would have to be a mechanism for accounting for the extra volume.

Eventually, the frequency would fall to such a low level the radiation would be absorbed into the fabric of space.

This would make far more sense of a static universe without any expansion or a big bang with inflation.

The Cosmic Microwave Background Radiation

I hypothesise, that what we see and refer to as the CMBR, is in fact the point at which electromagnetic radiation reaches saturation point. The different patterns in the CMB may be due to the varying density of galaxies in the universe in front of and beyond the CMB and the electromagnetic radiation being emitted from them.

Looked at logically, if we consider how many rays of electromagnetic radiation are travelling towards us when we look out into space, emanating from the billions of galaxies that exist over the distance of 13.8 billion light years and beyond, there must be a point of saturation, which is what we see as the CMB.

Really, the phenomenon referred to as the CMBR, ought to be referred to as the Electromagnetic Radiation Saturation Point (EMR-SP).

If we look into space in any direction, we will observe light or electromagnetic radiation emanating from countless galaxies at ever increasing distances. With our present equipment we can detect back a given distance, which at this time is around 14 billion years, where we interpret what we see as the CMB. However, I contend that what we see is actually the point where the electromagnetic radiation we are detecting, reaches saturation point for the equipment we have.

Fabric of Space

I will just give a very brief description of this, as it may help to make sense of this hypothesis. It is explained more fully in an hypothesis I have for the composition of the atom.

It is proposed that there exists, just two incredibly small monopole particles that make up everything that exists in the universe. One is a negatively charged monopole particle called a 'Harveytron', which fills every available empty space throughout the universe in a cloud, the 'Harveytron Cloud'. This cloud is the medium in which electromagnetic radiation is able to travel through space.

It is also the 'Dark matter', and the negative force of repulsion it produces, is the 'Dark Energy'.

This dark energy is also one of the two components of gravity, which I propose is a force of both attraction, and repulsion. It is also the force that keeps galaxies and the bodies contained in them, in their relative positions within the universe. It is also what stops galaxies from flying apart.

The other is a corresponding positively charged monopole particle called a 'Dannytron', which in combination with the 'Harveytrons', make up all of the atomic nuclei in the universe, and as such, all of the physical nuclear matter throughout the universe. All the other particles making up the standard model table of particles, including the electron, are composites of these two particles.

Creation of matter.

It is hypothesised that the matter that makes up all of the nuclear matter in the universe such as the planets etc, is produced by the combination of the two monopole particles I describe.

It is universally agreed that matter is formed as plasma cools. It is proposed that the early process of matter formation, is the emergence of the two monopole particles that I propose, with an overall massive excess of the negatively charged monopole 'Harveytron' particles. It is possible, that this process may be an endothermic reaction.

As soon as a positively charged monopole 'Dannytron' particle forms, it is immediately encompassed by the negatively charged monopole 'Harveytron' particles. It will be realised that it is not possible for the positively charged particles to exist on their own, whereas the negatively charged particles do.

If the process of matter formation is endothermic, absorbing heat in the process, thermal equilibrium is explained throughout the universe, without the need for 'Causal contact', and solves the 'Horizon problem'. Also, as I contend that the universe has always existed, thermal equilibrium is a constant with only minor hotspots, in relative terms, due to local events such as supernova and the like.

Cosmic Nucleosynthesis

It is claimed that the creation of deuterium and helium, along with trace amounts of lithium and beryllium, in the first 225 seconds after the big bang is proof of the big bang, because in the first moments of the universe, the temperature was billions of degrees. This was so hot that atoms were entirely ionised.

However, I claim that in a nonexpanding infinite universe, violent events such as supernova and the like, happen regularly throughout the universe creating the conditions for cosmic nucleosynthesis to occur, without the need for the big bang.

If it is accepted that the universe has always existed, and was not born all at once in the big bang, most of the problems explaining things such as the 'Horizon problem' and the like cease to be a problem. What has caused most of the problems in understanding cosmology, has come from a misconception of the 'CMB', and believing that beyond around 13-14 billion light years, there only existed hot plasma.

We know that throughout the universe, violent events, such as supernova exists, and matter is transformed into a plasma, also it is said to be consumed by black holes, only to be born again in stellar nurseries. This is a perpetual cyclic process, where all matter is not formed in one big event, rather evenly throughout the existing universe.

Antimatter

I believe that the first stage of matter formation produces the 'Antimatter', 'Dark Matter', and 'Dark Energy'.

The 'Antimatter' is the newly combined positively charged monopole particles encompassed with negatively charged monopole particles. This is the embryonic form of Hydrogen, prior to further combinations until enough positive and negatively charged particles cluster together to form a hydrogen nucleus. At this point the antimatter ceases to exist, as this now is recognised as a hydrogen atom.

I would hypothesise that the transition from antimatter to baryonic matter, is where the number of positive particles clustering together, become enough to allow a quanta of negative particles, or the energy they contain to be able to be received or emitted from what would be determined as an atom. Prior to this, 'Antimatter', which I contend is embryonic Hydrogen, does not have enough mass, and in particular the positively charged 'Dannytron' particles to be able to contain a quanta of energy.

This easily explains phenomena such as the Muonium. This is just what I have described above, and is likely to be just one of many clusters of the two particles that I propose starting to form nuclear matter, before they are big enough to have enough charge to allow them to eject a photon, and thus have an electron as described in the standard model associated with them.

Dark Matter

The 'Dark Matter' is the negatively charged monopole particles on mass in a cloud filling every available empty space throughout the universe trying to repel each other in every direction. Once all of the positively charged monopole 'Dannytron' particles have been absorbed following further cooling, there will only exist the negatively charged monopole 'Harveytron' particles free, in the 'Dark Matter' 'Harveytron Cloud'. The different elements being defined by the amount of positively charged monopole 'Dannytron' particles contained in the nucleus of each atom. The light elements being formed prior to star formation, and the heavier elements being produced within stars over time.

Dark Energy

The 'Dark Energy', is the repulsive force created by the negatively charged monopole 'Harveytron' particles trying to repel each other in every direction.

Electromagnetic radiation saturation

This was briefly explained earlier. If we consider how many rays of electromagnetic radiation we see when we look back through space, emanating from the vast number of stars and galaxies over that distance and time, there may come a point where the saturation is so great, that they merge into one.

This could easily explain why we see the phenomenon we call the CMB. This would make sense why it is the same in every direction, no matter where we exist in the universe. It would also explain why the universe has been found to be homogenous in every direction. The distance we can see, may be proportional to the receptive area of the instruments we use to detect the radiation.

This fact, may explain why the 'James Webb' telescope, can see further back than the 'Hubble', due to its larger mirror, however, it may be that this is the ultimate distance we will ever be able to see. We also have to consider the fact that because the radiation is not coming from one direction, but from every direction, there will be a mass of radiation passing each other in every direction, which must have some influence on what we see and measure.

The fact that the CMB has variations in its texture, would most likely be explained by the fact that the density of galaxies throughout the universe would result in varying concentrations of light rays being perceived by the observer.

Thermal equilibrium

Since the big bang never happened, there wouldn't have been a massive amount of heat to dissipate throughout the universe. This explains why the universe is a constant temperature, throughout and completely eliminates the 'Horizon Problem'. Not only that, but the ridiculous notion of the big bang, where everything emanated from a singularity smaller than an atom, highlights the pitfalls that arise when mathematics are used at the expense of logic. We should ask the question, 'How is it possible for something as small as a singularity to contain enough heat to raise the temperature of all the billions of tons of matter that supposedly filled the early universe with a seething hot soup of a billion degrees C?' It really is nonsense.

Black Holes

It is my view that the current accepted understanding of these phenomena may be confused by preconceived mathematical calculations of the fate of dying stars.

It is stated that when stars die, they explode in a supernova, then collapse into themselves to form a neutron star creating a black hole of immense gravity. These assumptions I believe are derived purely from mathematical equations.

If looked at logically the description above does not make sense. Firstly, if the star exploded forcing all of its matter out into space, why would it collapse back in on itself? It is argued that this is due to gravity, however, I would argue that gravity is a property of matter, and in such an explosion, all of the matter from the dead star, would have been spread throughout the surrounding area, and would most likely combine with surrounding stellar masses.

It is stated that it is just the outer layer that explodes of, who thought that one up? If a star is hot enough to explode, it is likely that the core would be molten and the whole star would disintegrate and be blasted out into the surrounding universe. From common logic, it will be realised that since gravity is a property of matter, there will be no gravity left where the centre of the star had been, except for the 'Harveytron' cloud, which fills every empty space throughout the universe.

Depending on the size of the star that exploded and the speed at which it was rotating, will determine the size of the void left where the star had been, (A Black Hole).

This would result in a region of very little matter where the star had been. and due to the lack of aggregate mass, what debris that was left, would be forced out by centrifugal force, thereby forming a 'Black Hole' where the star had been.

Gravity within a Black Hole

In my model, I would contend that the gravity within the black hole would be just the same negative force of repulsion as that of the 'Dark Energy' resulting from the 'Dark Matter' 'Harveytron' cloud that I propose. It is very likely that within the black hole there would be very little or no Baryonic matter.

In essence, a black hole is just a vortex formed by the stellar matter revolving around it, the size of the black hole being dependent on how much matter is contained within that gravitational sphere.

Galaxy Birth and Formation

I hypothesise that this is how Galaxies are formed.

When a star explodes and forms a black hole the additional matter and associated gravity that is spread out in the region surrounding the black hole where the star was situated, forms an embryonic galaxy, which gradually grows as matter is gradually acquired due to the fact that gravitational matter has been forced further out from where the star was situated.

It would be very unlikely that all of the matter that was exploded out from the star would collapse back to form the same star because between every pieces of the exploded matter, there will be the negative force of repulsion of the 'Dark Energy' being produced by the 'Dark Matter' 'Harveytron cloud' forcing baryonic matter apart, depending on its mass and distance between each piece of matter. It is likely that small clumps would form dispersed throughout the area covered by the explosion. These small clumps will eventually evolve into embryonic stars as more matter is accreted over millions of years.

Galaxy Expansion

I do not believe that matter is sucked into a black hole. I believe that the only thing that exists in a black hole is the 'Harveytron Dark Matter Cloud'. We have to consider the fact that there are two sides to a black hole, so what happens on one side happens on the other.

Any baryonic matter that gets too near a black hole, including 'Antimatter', will be forced out to the edge of the black hole, slowly adding to the amount of matter and therefore gravity of this new galaxy. The size of the black hole at the centre, will be directly proportional to the amount of the positively charged 'Dannytron' particles contained in the nuclei of the baryonic matter in the galaxy, because this will determine how many of the negatively charged 'Harveytron' particle are held in the atoms contributing to the overall mass of the galaxy.

If a black hole had the immense gravity that the standard model proposes, the black hole would not exist, as any matter surrounding it, would be pulled to the centre and a new star would exist.

Light and all Electromagnetic Radiation, The Accretion Disc, and Hawking Radiation

Any light entering the black hole from either side, will be deflected out to the edge of the black hole, where it could cause a reaction similar to that of the 'Photoelectric Effect' thereby accounting for the 'Accretion Disc', and 'Hawking Radiation'.

This is why, light does not reflect out of the centre of the black hole. It is not that it can't escape the gravity of the black hole, it can, but from the perimeter of the black hole. It also will be realised that since there is no baryonic matter in the vortex of the black hole, there is nothing that light can reflect off of.

In my model, theoretically, you could pass through the centre of the black hole, where the centrifugal force should be zero, to reach the other side of the galaxy.

Conclusion

Apart from the absurdity of everything that exists in the universe emerging from a singularity smaller than an atom, and this singularity inflating to the size of the universe in a fraction of a billionth of a second, the standard model still can't explain 'Dark Matter', 'Dark Energy', what is the repulsive force field in the universe that stays at a constant pressure despite the supposed expansion of the universe, or 'Gravity'. It also can't reconcile the conflict between quantum mechanics, classical physics, and cosmology.

The fact that the CMB can be seen in every direction, casts serious doubts on the validity of the currently accepted model.

The 'Big Bang' theory, and the 'Relativistic Cosmological Model', can only be justified by the ridiculous idea of 'Cosmic Inflation', which defies logic, and violates all the laws of physics, including matter travelling through the universe at multiple times the speed of light. I contend that these facts, renders this theory disproved.

The two above hypotheses can explain logically many of the outstanding problems in physics and cosmology.

Hypothesis proffered by: Tony Norman Marsh